We Live Here

Harcourt
SCHOOL PUBLISHERS

Visit *The Learning Site!* www.harcourtschool.com

The Inuit

We live in a cold place. There is a lot of snow in this location.

Once we lived in homes made of ice. Now families live in houses made of wood.

We get food from the sea. It is an important resource.

The Tuareg

We live in a hot, dry place. Once we moved from place to place. We slept in tents. Now we live in villages.

Once we followed the animals. They looked for water, and we did, too. Now we ride some animals. They also give us milk. Some of us grow food.

The Yanomami

We live in the rain forest. It is hot and wet. All of the people in our village live in one house.

The natural resources of the rain
forest give us food. We hunt animals
in the forest. We fish in the river.
We grow food in our gardens.

Think and Respond

1. What is the weather like where the Inuit live?

2. How do the Tuareg use their animals?

3. Where do the Yanomami get their food?

4. How do the different groups dress for where they live?

5. How have the Tuareg changed over time?

Activity

Make a book about the place where you live.

Photo Credits

Cover Layne Kennedy/Corbis; P.2 Galen Rowell/Corbis; P.3 Alain Le Garsmeur/Corbis; P.4 Tiziana and Gianni Baldizzone/Corb
P.5 G. Rossenbach/zefa/Corbis; P.6 Corbis SYGMA; P.7 Victor Englebert/Time Life Pictures/Getty Images.

Illustration Credits

P.2 Bill Melvin; P.4 Bill Melvin, P.6 Bill Melvin, Back Cover Bill Melvin.